# Singapore Math Kindergarten

# Help Your Child Develop Early Math Skills.

Math skills are vital in early education, and are extremely important for logical thinking required by multiple school subjects.

These math activities for kids make learning math fun and engaging! you will find lots of hands-on and playful ways for kids to learn math concepts. They are perfect for math centers, small group instruction or homeschool.

Using games to improve basic math skills is a great learning strategy for kids. This book contains cool math games designed to keep kids engaged while learning math skills.

# Table of Contents

## Addition

Count, add the pictures and write your answer in the place holder.

 +  = ☐

 +

 +  = ☐

 +

 +  = ☐

 +

## Addition

Count, add the pictures and write your answer in the place holder.

 +  =

 +

 +

 +

 +  =

 +

Count, add the pictures and write your answer in the place holder.

 +  =

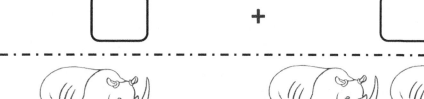

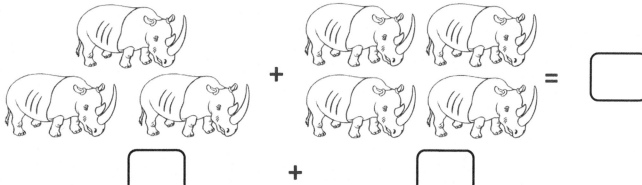

## Addition

Count, add the pictures and write your answer in the place holder.

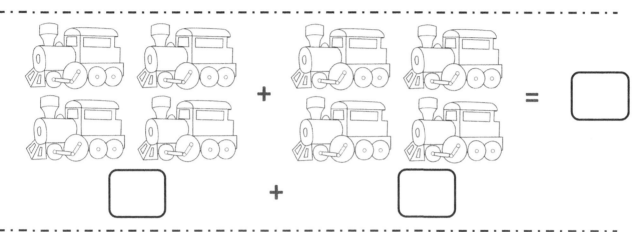

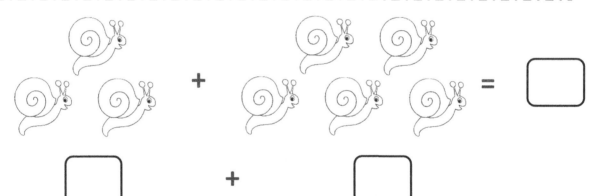

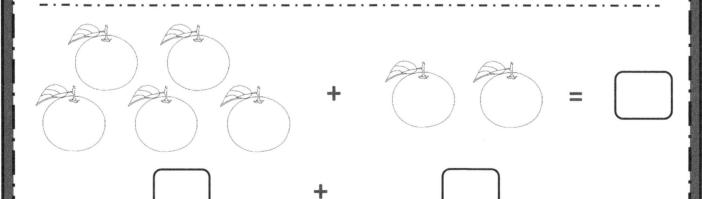

# Addition

Count, add the pictures and write your answer in the place holder.

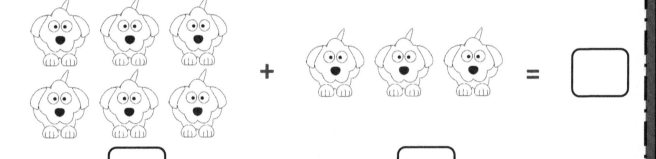

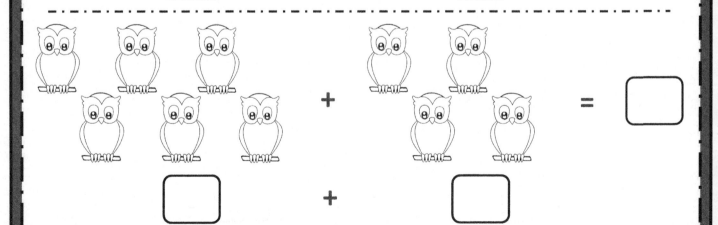

# Addition

Count, add the pictures and write your answer in the place holder.

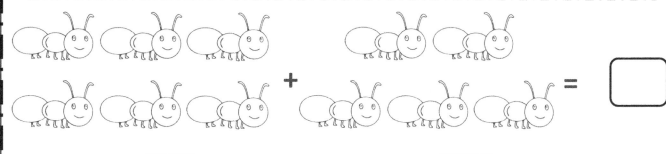

$+$ $=$ [ ]

[ ] $+$ [ ]

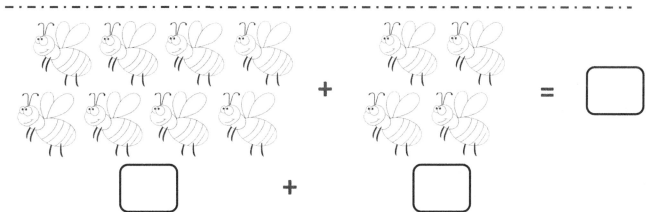

[ ] $+$ [ ]

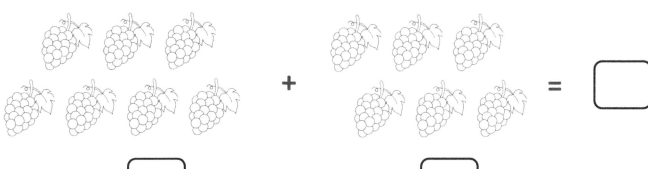

[ ] $+$ [ ]

# Addition

Count, add the pictures and write your answer in the place holder.

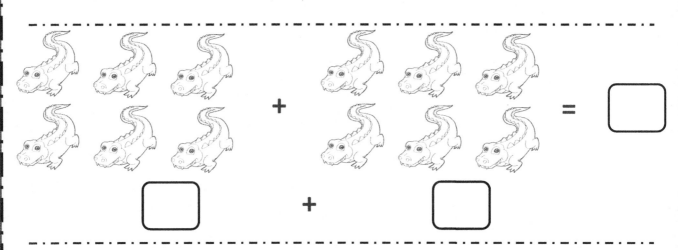

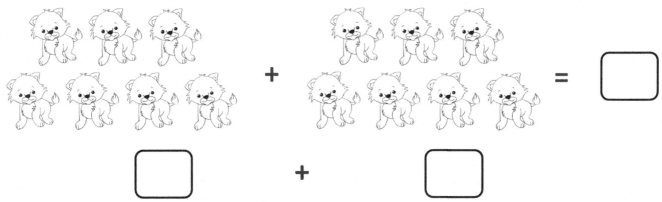

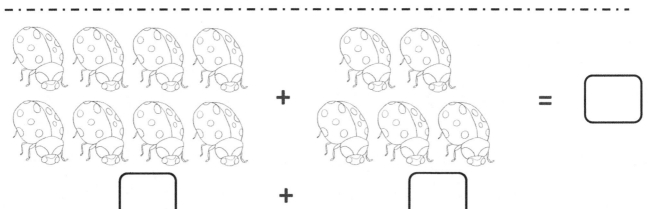

# Addition

Count, add the pictures and write your answer in the place holder.

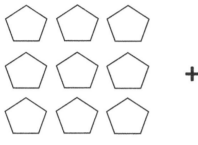 + 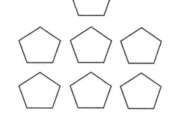 =

 +

 =

 +

 +

Count, add the pictures and write your answer in the place holder.

☆☆☆
☆☆☆   +   ☆☆☆
☆☆☆       ☆☆☆
          ☆☆☆     =   [ ]

[ ]   +   [ ]

△
△△△       △
△△△   +   △     =   [ ]
△△△       △
          △

[ ]   +   [ ]

((((( (((((
((((( (((((   +   ((((( (((((   =   [ ]

   +

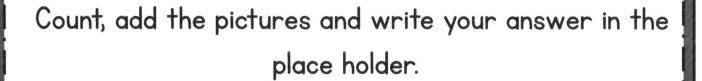

# Addition

Count, add the pictures and write your answer in the
place holder.

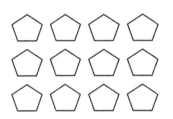    +        =    □

□    +    □

□    +    ☼☼☼☼    =    □

□    +    □

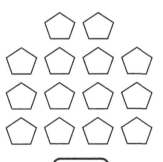    +    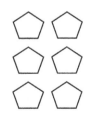    =    □

□    +    □

Count, add the pictures and write your answer in the place holder.

 +  = ☐

☐ + ☐

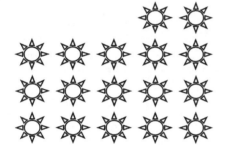 + ☐ = ☐

☐ + ☐

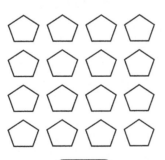 + 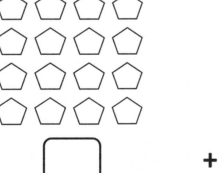 = ☐

☐ + ☐

# Addition

Find the sum.

| | | |
|---|---|---|
| 3<br>+ 2<br>___ | 4<br>+ 1<br>___ | 1<br>+ 2<br>___ |
| 5<br>+ 1<br>___ | 0<br>+ 4<br>___ | 2<br>+ 2<br>___ |
| 4<br>+ 3<br>___ | 3<br>+ 3<br>___ | 2<br>+ 4<br>___ |

# Addition

Find the sum.

$$5 + 2 = $$

$$4 + 4 = $$

$$1 + 6 = $$

$$3 + 6 = $$

$$5 + 5 = $$

$$4 + 5 = $$

$$7 + 2 = $$

$$3 + 5 = $$

$$6 + 4 = $$

# Addition

Find the sum.

| | | |
|---|---|---|
| 2<br>+ 8 | 6<br>+ 6 | 1<br>+ 9 |
| 7<br>+ 4 | 3<br>+ 7 | 8<br>+ 3 |
| 6<br>+ 5 | 2<br>+ 9 | 5<br>+ 7 |

# Addition

Find the sum.

| | | |
|---|---|---|
| 7<br>+ 7 | 8<br>+ 5 | 9<br>+ 5 |
| 6<br>+ 8 | 8<br>+ 8 | 7<br>+ 9 |
| 7<br>+ 8 | 8<br>+ 9 | 9<br>+ 9 |

# Addition

The numbers in the circles added together makes the number in the linking rectangle. Find the missing numbers in this puzzle.

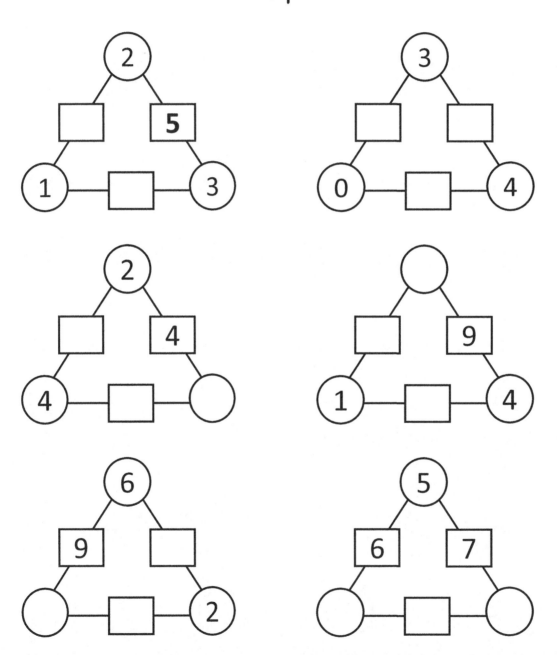

# Addition

The numbers in the circles added together makes the number in the linking rectangle. Find the missing numbers in this puzzle.

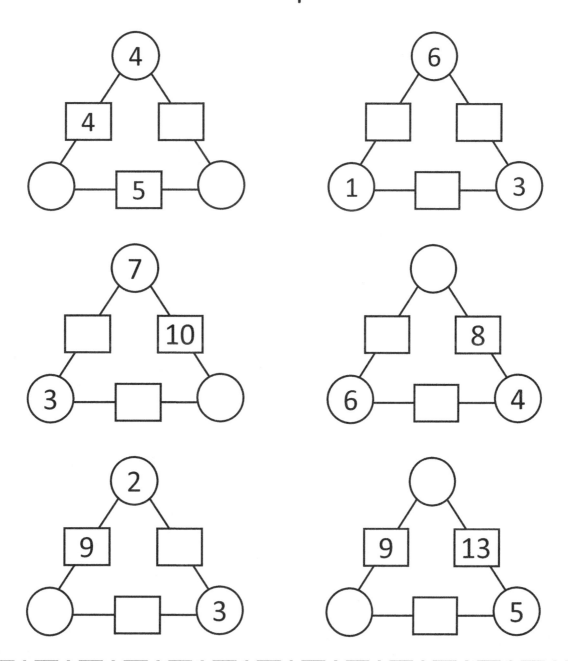

The numbers in the circles added together makes the number in the linking rectangle. Find the missing numbers in this puzzle.

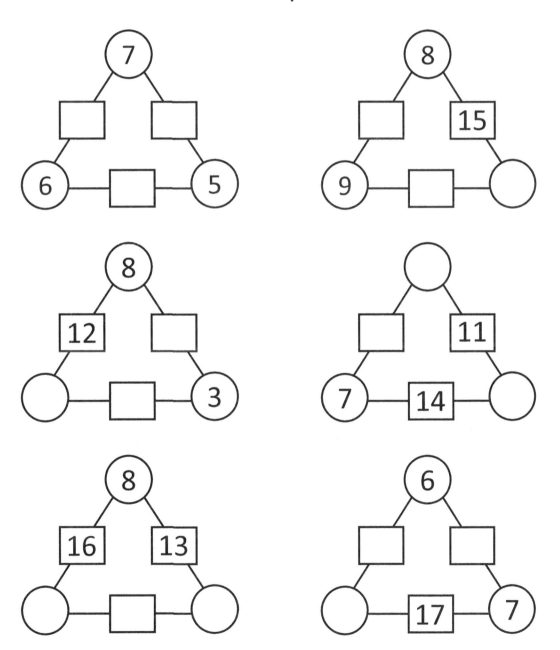

21

# Addition

Think of four ways to make a group of **8** elephants.

*Example* {

5 + 3 = 8

6 + 2 = 8

☐ + ☐ = 8

☐ + ☐ = 8

# Addition

Think of four ways to make a group of **9** cars.

☐ + ☐ = 9

☐ + ☐ = 9

☐ + ☐ = 9

☐ + ☐ = 9

## Addition

Think of four ways to make a group of 10 foxes.

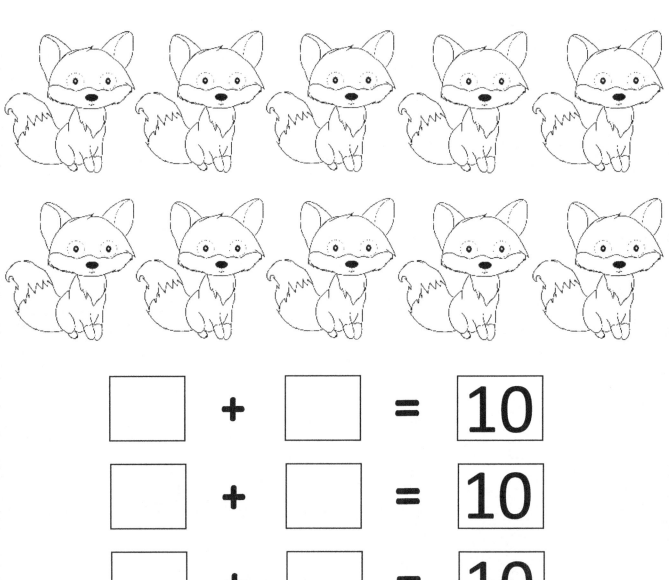

$$\boxed{\phantom{0}} + \boxed{\phantom{0}} = \boxed{10}$$

$$\boxed{\phantom{0}} + \boxed{\phantom{0}} = \boxed{10}$$

$$\boxed{\phantom{0}} + \boxed{\phantom{0}} = \boxed{10}$$

$$\boxed{\phantom{0}} + \boxed{\phantom{0}} = \boxed{10}$$

## Addition

Think of four ways to make a group of 11 owls.

☐ + ☐ = 11

☐ + ☐ = 11

☐ + ☐ = 11

☐ + ☐ = 11

## Addition

Read and solve the problems.

1. Adam has three chickens and five ducks. How many fowls does he have?

2. Joe has five popsicle sticks. I have four popsicle sticks. What is the sum of our popsicle sticks?

3. Gino and Sam are playing a computer game. Gino scores eight points and Sam scores seven points. How many points do they score altogether?

# Addition

Read and solve the problems.

1. There are ten birds on the fence. Nine more birds land on the fence. How many birds are on the fence?

2. Anita went to the grocery store. She bought 10 packs of cookies and 10 packs of noodles. How many packs of groceries did she buy in all?

3. There are three rows of seats in each of first class and business class and there are thirteen rows of seats in economy class. How many rows of seats are there on the plane?

## Addition

Read and solve the problems.

1. Sara played tag with 13 kids on Monday. She played tag with 5 kids on Tuesday and 2 kids on Wednesday. How many kids did she play with altogether?

2. Linda has 5 apples. Julia gave her 7 more. She needs 14 apples to make a pie. Does she have enough to make a pie?

3. There is a flower shop at downtown which is having a sale on roses. In the display, there are 8 white roses, 8 pink roses and 3 red roses. How many roses are there in total?

# Addition

Read and solve the problems.

1. Catherine bought 2 candies for 7 cents and 3 bubble gums for 12 cents each. How much did she spend in all?

2. Sara made 7 Rice Krispie Treats. She used 9 large marshmallows and 12 mini marshmallows. How many marshmallows did she use altogether?

3. Lucy ate 2 apples every hour. How many apples had she eaten at the end of 3 hours?

# Addition

Read and solve the problems.

1. Emma saw four bugs eat five flowers each. How many flowers total did the bugs eat?

2. Isabella bought 3 pizzas. Each pizza had 6 slices. How many total slices of pizza did she have?

3. Benjamin read 2 books per day. How many books did he read in one week?

# Subtraction

Count, subtract and write your answer in the place holder.

---

$= \boxed{1}$

$\boxed{3} - \boxed{2}$

---

$= \boxed{\phantom{0}}$

$\boxed{2} - \boxed{1}$

---

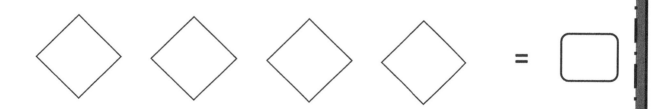

$= \boxed{\phantom{0}}$

$\boxed{4} - \boxed{2}$

## Subtraction

Count, subtract and write your answer in the place holder.

| 5 | - | 1 | = | ☐ |

| 4 | - | 4 | = | ☐ |

| 5 | - | 3 | = | ☐ |

## Subtraction

Count, subtract and write your answer in the place holder.

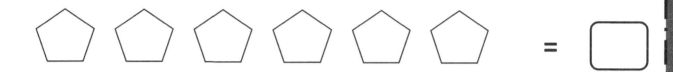  =  [ ]

| 6 | - | 3 |

(((((( ( (  =  [ ]

| 7 | - | 2 |

♡ ♡ ♡ ♡ ♡ ♡  =  [ ]

| 6 | - | 5 |

## Subtraction

Count, subtract and write your answer in the place holder.

☆ ☆ ☆ ☆ ☆ ☆ ☆ = ▢

7 - 3

△ △ △ △ △ △ △ △ = ▢

8 - 2

☼ ☼ ☼ ☼
☼ ☼ ☼ ☼ ☼ = ▢

9 - 5

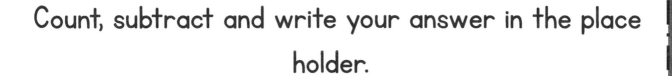

# Subtraction

Count, subtract and write your answer in the place holder.

| 8 | - | 4 | = | ☐ |

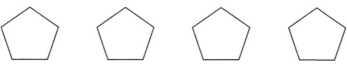

| 9 | - | 7 | = | ☐ |

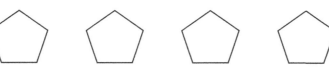

| 10 | - | 5 | = | ☐ |

# Subtraction

Count, subtract and write your answer in the place holder.

12    -    5    = ☐

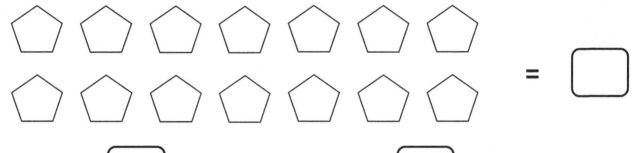

14    -    8    = ☐

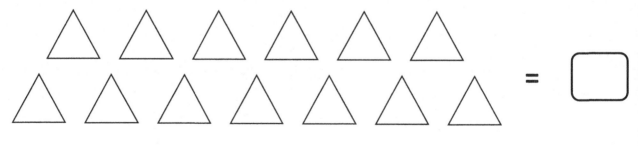

13    -    4    = ☐

# Subtraction

Count, subtract and write your answer in the place holder.

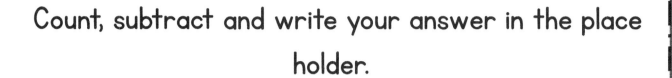

| 15 | - | 6 | = | ☐ |

| 17 | - | 9 | = | ☐ |

| 20 | - | 13 | = | ☐ |

# Subtraction

Think of three ways to make a group of **5** camels.

*Example* ➡ | 7 | - | 2 | = | 5 |

| | - | | = | 5 |

| | - | | = | 5 |

# Subtraction

Think of three ways to make a group of **6 goats.**

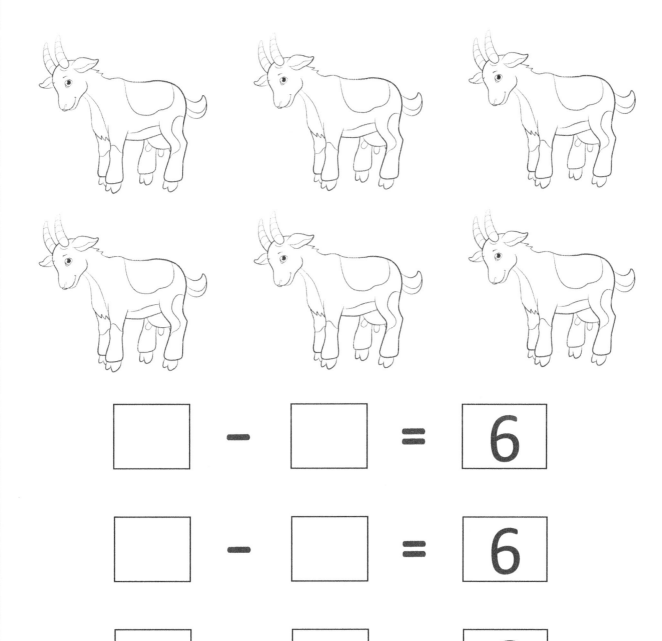

$$\boxed{\phantom{6}} - \boxed{\phantom{6}} = \boxed{6}$$

$$\boxed{\phantom{6}} - \boxed{\phantom{6}} = \boxed{6}$$

$$\boxed{\phantom{6}} - \boxed{\phantom{6}} = \boxed{6}$$

# Subtraction

Think of three ways to make a group of **7 mice.**

☐ − ☐ = 7

☐ − ☐ = 7

☐ − ☐ = 7

# Subtraction

Think of three ways to make a group of 8 snakes.

$$\boxed{\phantom{8}} - \boxed{\phantom{8}} = \boxed{8}$$

$$\boxed{\phantom{8}} - \boxed{\phantom{8}} = \boxed{8}$$

$$\boxed{\phantom{8}} - \boxed{\phantom{8}} = \boxed{8}$$

Find the difference.

| | | |
|---|---|---|
| $\begin{array}{r} 4 \\ -\ 2 \\ \hline \end{array}$ | $\begin{array}{r} 5 \\ -\ 1 \\ \hline \end{array}$ | $\begin{array}{r} 3 \\ -\ 3 \\ \hline \end{array}$ |
| $\begin{array}{r} 5 \\ -\ 2 \\ \hline \end{array}$ | $\begin{array}{r} 4 \\ -\ 3 \\ \hline \end{array}$ | $\begin{array}{r} 6 \\ -\ 2 \\ \hline \end{array}$ |
| $\begin{array}{r} 4 \\ -\ 1 \\ \hline \end{array}$ | $\begin{array}{r} 5 \\ -\ 3 \\ \hline \end{array}$ | $\begin{array}{r} 6 \\ -\ 3 \\ \hline \end{array}$ |

# Subtraction

## Find the difference.

$$\begin{array}{r} 6 \\ -\ 1 \\ \hline \end{array}$$
$$\begin{array}{r} 4 \\ -\ 4 \\ \hline \end{array}$$
$$\begin{array}{r} 5 \\ -\ 4 \\ \hline \end{array}$$

$$\begin{array}{r} 6 \\ -\ 4 \\ \hline \end{array}$$
$$\begin{array}{r} 7 \\ -\ 3 \\ \hline \end{array}$$
$$\begin{array}{r} 8 \\ -\ 3 \\ \hline \end{array}$$

$$\begin{array}{r} 7 \\ -\ 1 \\ \hline \end{array}$$
$$\begin{array}{r} 6 \\ -\ 5 \\ \hline \end{array}$$
$$\begin{array}{r} 7 \\ -\ 2 \\ \hline \end{array}$$

# Subtraction

Find the difference.

| | | |
|---|---|---|
| 7<br>− 4 | 8<br>− 4 | 9<br>− 5 |
| 8<br>− 6 | 9<br>− 7 | 7<br>− 5 |
| 9<br>− 6 | 8<br>− 1 | 9<br>− 3 |

# Subtraction

Read and solve the problems.

1. William has seven marbles. Then William gives two marbles to Gino. How many marbles does William have left now?

2. Sam has nine oranges. Then Sam gives four oranges to Lucy. How many oranges does Sam have left now?

3. James has three fewer pens than Adam. Adam has ten pens. How many pens does James have?

# Subtraction

Read and solve the problems.

1. Javana has four fewer balls than Lucy. Lucy has eleven balls. How many balls does Javana have?

2. There are 16 pens are in the backpack. 9 are red and the rest are green. How many pens are green?

3. Joe has 10 fewer cucumbers than Jen. Jen has 17 cucumbers. How many cucumbers does Joe have?

## Subtraction

Read and solve the problems.

1. There are 18 bird families and 14 mammals families living near the mountain. If 12 bird families flew away for winter, how many bird families were left near the mountain?

2. James needs at least 20 points to go to level 5 in a video game. He has only 7 points in level 4. How many more points does he need to qualify for level 5?

3. Joe had 20 marbles. He gave 7 to Michael and 7 to William. How many does he have left?

## Subtraction

Read and solve the problems.

1. Dora's pencil is 20 inches long. If she sharpens four inches off on Wednesday and two inches on Thursday and five inches on Friday, how long will her pencil be then?

2. Elijah found 4 small whiteboards and 7 boxes of markers for coaches to use. Oliver said they need a total of 13 whiteboards and 10 boxes of markers. How many more whiteboards and boxes of markers were needed?

3. James has a box of toy vehicles. There are 19 cars, 30 trucks and 42 emergency vehicles. 10 cars are blue and 3 are white. The rest are red. How many red cars are there in the box?

# Comparing Numbers

## Greater Than ( > )

Three is greater than two :  $\boxed{3}$  $>$  $\boxed{2}$

## Less Than ( < )

One is Less than two :  $\boxed{1}$  $<$  $\boxed{2}$

## Equal ( = )

Two equals two :  $\boxed{2}$  $=$  $\boxed{2}$

# Comparing Numbers

Compare with >,< or =.

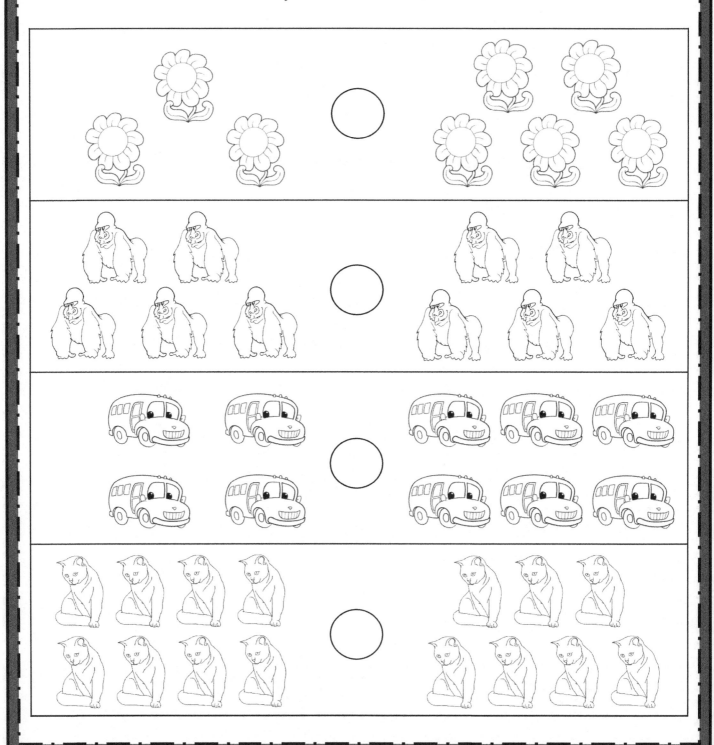

# Comparing Numbers

Compare with >,< or =.

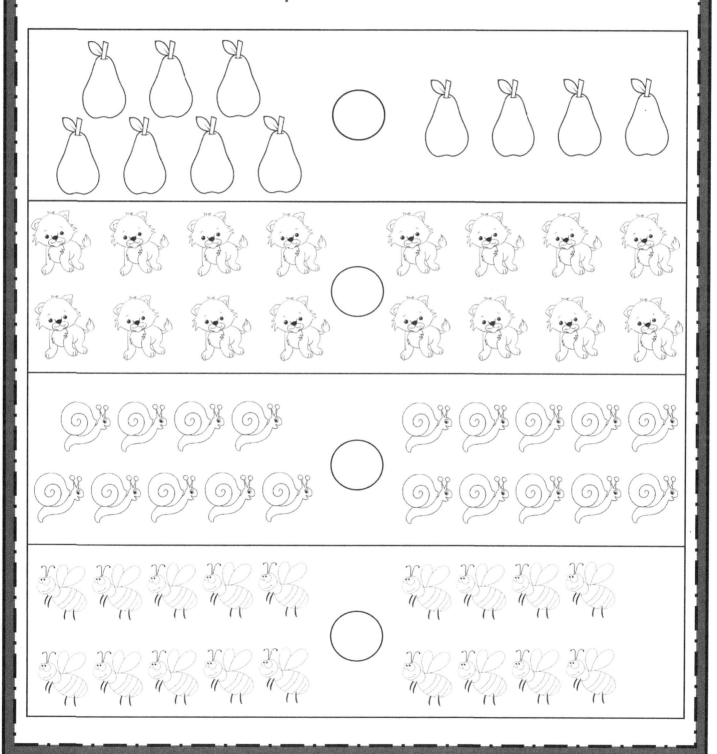

# Comparing Numbers

Compare with >,< or =.

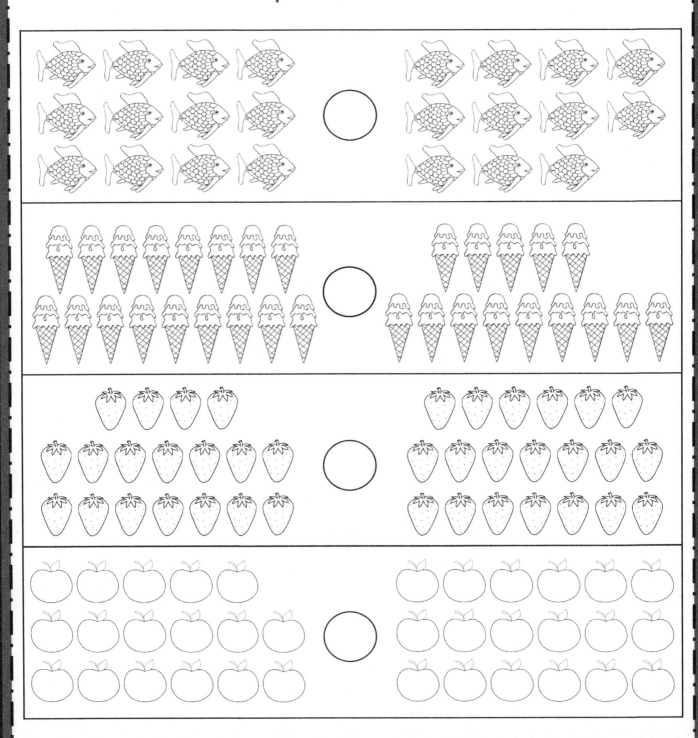

# Comparing Numbers

Compare with >,< or =.

| 4 | ◯ | 1 | | 8 | ◯ | 3 |

| 5 | ◯ | 9 | | 9 | ◯ | 10 |

| 7 | ◯ | 7 | | 6 | ◯ | 8 |

| 9 | ◯ | 2 | | 6 | ◯ | 4 |

| 10 | ◯ | 8 | | 7 | ◯ | 10 |

# Comparing Numbers

Compare with >,< or =.

14 ◯ 9          11 ◯ 10

15 ◯ 18          8 ◯ 12

16 ◯ 20          19 ◯ 5

17 ◯ 16          18 ◯ 20

19 ◯ 19          12 ◯ 14

# Comparing Numbers

## Order from least to greatest.

3 2 4 ➡ ☐ ☐ ☐

8 1 5 ➡ ☐ ☐ ☐

4 8 9 ➡ ☐ ☐ ☐

13 8 12 ➡ ☐ ☐ ☐

10 13 9 ➡ ☐ ☐ ☐

18 19 14 ➡ ☐ ☐ ☐

3 10 7 ➡ ☐ ☐ ☐

# Comparing Numbers

## Order from least to greatest.

3  8  2  4  ➡  ☐ ☐ ☐ ☐

9  3  1  8  ➡  ☐ ☐ ☐ ☐

6  10  7  5  ➡  ☐ ☐ ☐ ☐

4  8  13  11  ➡  ☐ ☐ ☐ ☐

15  10  16  18  ➡  ☐ ☐ ☐ ☐

10  8  17  9  ➡  ☐ ☐ ☐ ☐

12  14  16  18  ➡  ☐ ☐ ☐ ☐

# Comparing Numbers

Order from least to greatest.

3  8  10  2  6  ➡  ▢ ▢ ▢ ▢ ▢

10  5  9  7  1  ➡  ▢ ▢ ▢ ▢ ▢

4  6  8  7  9  ➡  ▢ ▢ ▢ ▢ ▢

3  10  2  13  11  ➡  ▢ ▢ ▢ ▢ ▢

16  5  18  17  13  ➡  ▢ ▢ ▢ ▢ ▢

11  14  19  9  7  ➡  ▢ ▢ ▢ ▢ ▢

18  12  13  16  19  ➡  ▢ ▢ ▢ ▢ ▢

57

# Measurement

Answer the question by coloring the correct picture.

Which is longer?

Which is shorter?

Which is longer?

Which is shorter?

Which is not the longer and not the shorter ?

Which is not the longer and not the shorter ?

# Measurement

Wich of these objects is long , longer and longest?

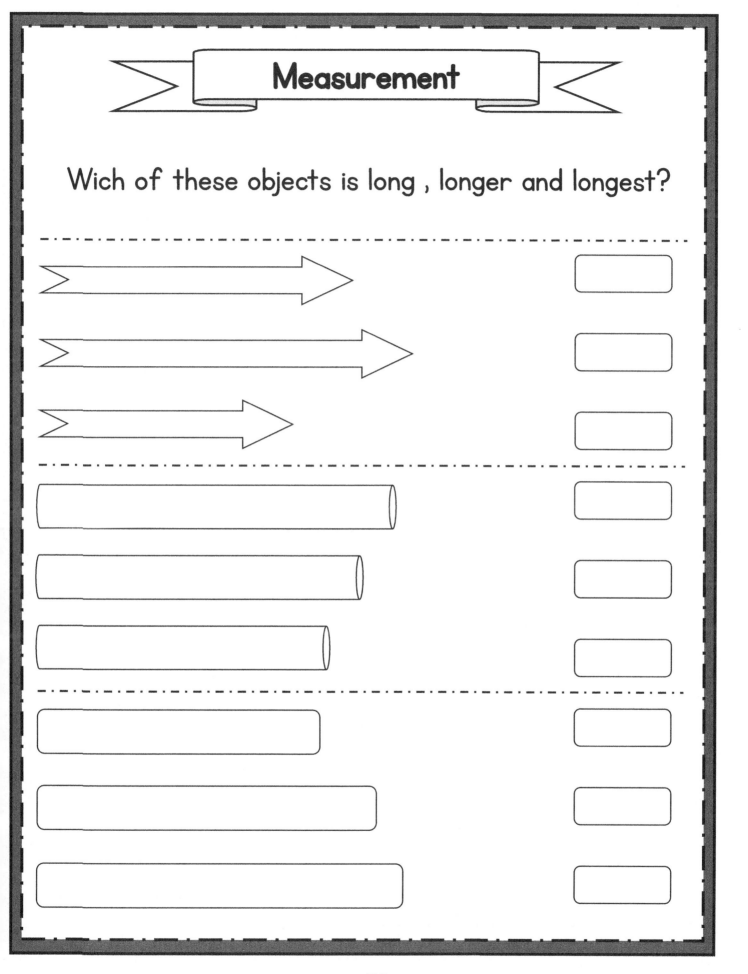

# Measurement

Measure and write the number of blocks for each object.

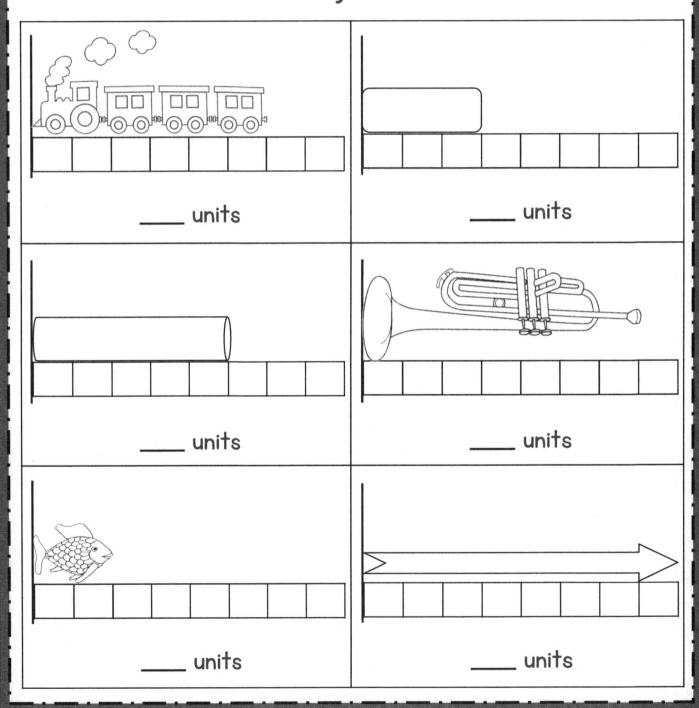

____ units

____ units

____ units

____ units

____ units

____ units

# Measurement

Answer the question by coloring the correct picture.

Color the heaviest animal.

Color the lightest fruits.

Color the animal that is not the heaviest and not the lightest.

# Measurement

Answer the question by coloring the correct picture.

Color the heaviest object.

  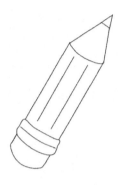

Color the lightest vehicle.

Color the object that is not the heaviest and not the lightest.

# Measurement

Circle the correct word.

lighter

as heavy as

The rooster is _____ Than the iron.

heavier

as heavy as

The bull is _____ Than the dog.

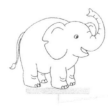 

heavier

lighter

The lion is _____ Than the elephant.

## Measurement

Circle the correct word.

**heavier**

**lighter**

The pumpkin is _____ Than the apple.

**heavier**

**as heavy as**

The cake is _____ Than the owl.

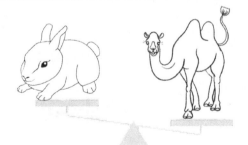

**heavier**

**lighter**

The rabbit is _____ Than the camel.

# Time

Draw the time shown on each clock.

| 2:00 | 10:00 | 3:30 | 5:00 |

| 4:30 | 6:00 | 8:30 | 7:00 |

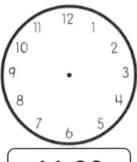

| 11:30 | 1:30 | 9:00 | 12:30 |

# Time

Draw the time shown on each clock.

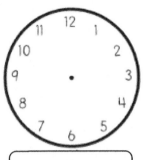

| 4:45 |

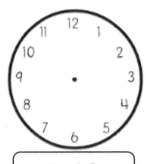

| 7:10 |

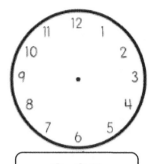

| 8:25 |

| 2:50 |

| 11:05 |

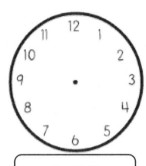

| 1:55 |

| 9:20 |

| 3:35 |

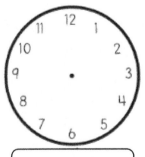

| 5:40 |

| 12:15 |

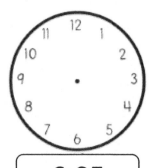

| 6:05 |

| 10:25 |

# Time

Draw the time shown on each clock.

7:50

8:20

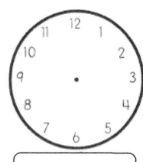

1:10

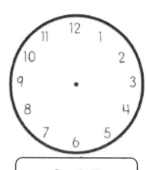

3:15

6:35

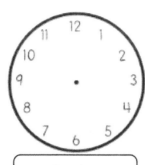

9:45

11:20

5:05

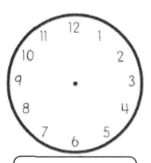

12:35

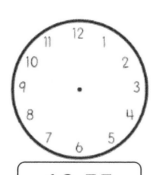

10:55

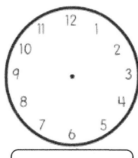

2:15

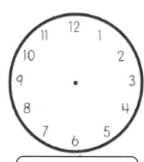

4:10

# Time

Circle the best estimate of the time needed for each activity.

Taking school bus to school.

Minutes       Hours       Days

Going camping.

Minutes       Hours       Days

Go on for a fishing trip.

3 years       3 days       3 minutes

Taking a morning jog.

20 months       20 minutes       20 seconds

Counting from I to 30.

30 minutes       30 seconds       30 days

# Time

Circle the best estimate of the time needed for each activity.

Preparing dinner.

Months          Years          Minutes

Taking a picture.

Years          Seconds          Months

Brushing your teeth.

1 week          1 minute          1 year

Building a new bridge.

3 days          3 months          3 minutes

Reading.

Seconds          Minutes          Weeks

# Time

Draw the clock hands to show the time it was or will be.

What time will it be in 2 hour 0 minutes?

What time will it be in 4 hours 0 minutes?

What time was it 2 hours 0 minutes ago?

What time was it 3 hours 0 minutes ago?

What time will it be in 4 hours 20 minutes?

What time was it 1 hours 50 minutes ago?

70

# Time

Draw the clock hands to show the time it was or will be.

What time will it be in 4 hour 45 minutes?

What time will it be in 3 hours 15 minutes?

What time was it 1 hours 25 minutes ago?

What time was it 1 hours 30 minutes ago?

What time will it be in 5 hours 35 minutes?

What time was it 4 hours 40 minutes ago?

# Time

## Does this happen in the a.m. or p.m.?

| | | |
|---|---|---|
| Postman delivering mail in the morning | ➡ | **A. M. / P. M.** |
| Playing toys after dinner | ➡ | **A. M. / P. M.** |
| Stargazing | ➡ | **A. M. / P. M.** |
| Birthday party on a Saturday morning | ➡ | **A. M. / P. M.** |
| Rooster crowing at dawn | ➡ | **A. M. / P. M.** |
| Riding bike after lunch | ➡ | **A. M. / P. M.** |
| Taking school bus to school | ➡ | **A. M. / P. M.** |

# Geometry

Draw a line between a shape and its name.

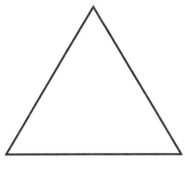 Rectangle

 Cercle

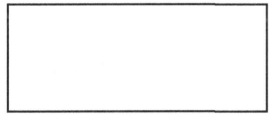 Square

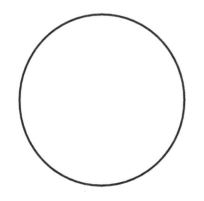 Triangle

# Geometry

Draw a line between a shape and its name.

Pentagon

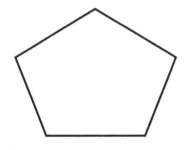

Rhombus

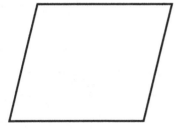

Trapezoid

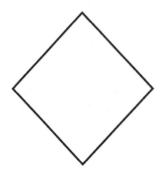

Parallelogram

# Geometry

Draw a line between a shape and its name.

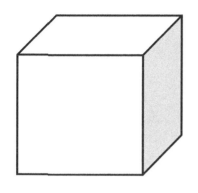
Cylinder

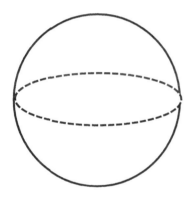
Cone

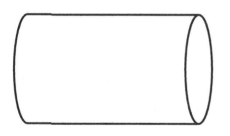
Cube

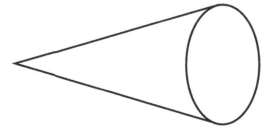
Sphere

# Geometry

Trace and color triangles.

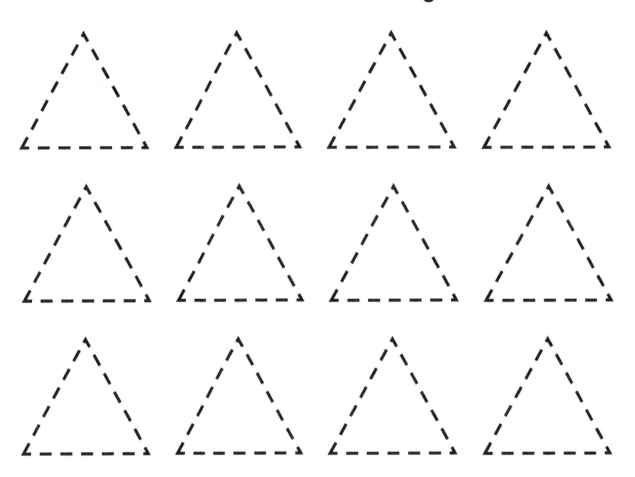

Count how many sides and vertices for triangle.

| Number of Sides | Number of Vertices |
| --- | --- |
|  |  |

# Geometry

Trace and color cercles.

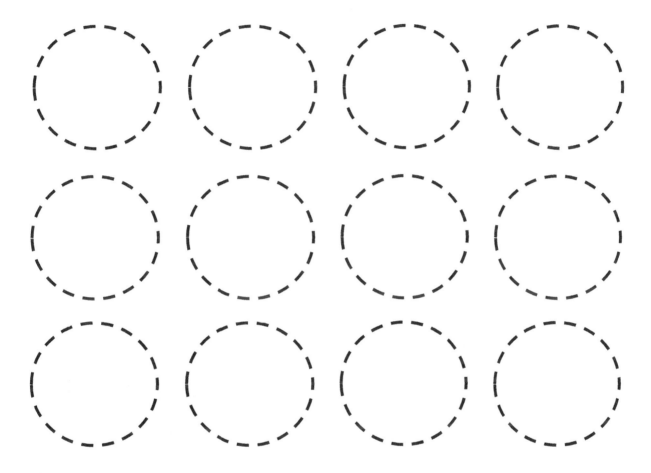

Count how many sides and vertices for cercle.

| Number of Sides | Number of Vertices |
| --- | --- |
|  |  |

# Geometry

Trace and color squares.

Count how many sides and vertices for square.

| Number of Sides | Number of Vertices |
|---|---|
|  |  |

# Geometry

Trace and color rectangles.

Count how many sides and vertices for rectangle.

| Number of Sides | Number of Vertices |
|---|---|
|  |  |

# Geometry

Trace and color parallelograms.

Count how many sides and vertices for parallelogram.

| Number of Sides | Number of Vertices |
| --- | --- |
|  |  |

# Geometry

Trace and color rhombuses.

Count how many sides and vertices for rhombuse.

| Number of Sides | Number of Vertices |
| --- | --- |
|  |  |

Trace and color trapezoids.

Count how many sides and vertices for trapezoid.

| Number of Sides | Number of Vertices |
| --- | --- |
|  |  |

# Geometry

Trace and color Ellipses.

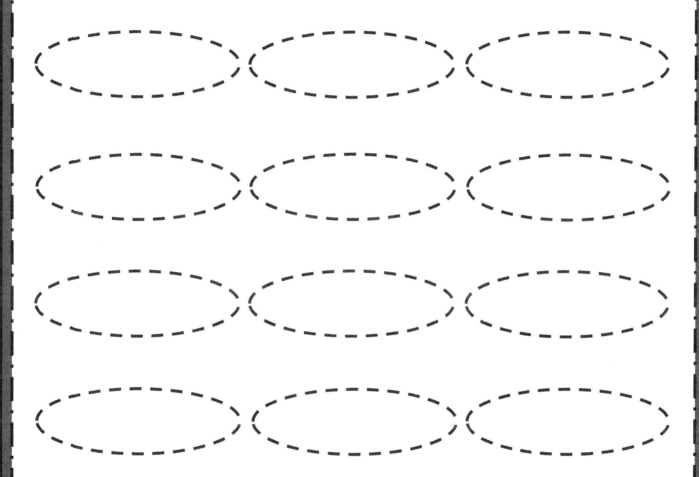

Count how many sides and vertices for Ellipse.

| Number of Sides | Number of Vertices |
| --- | --- |
|  |  |

Trace and color pentagons.

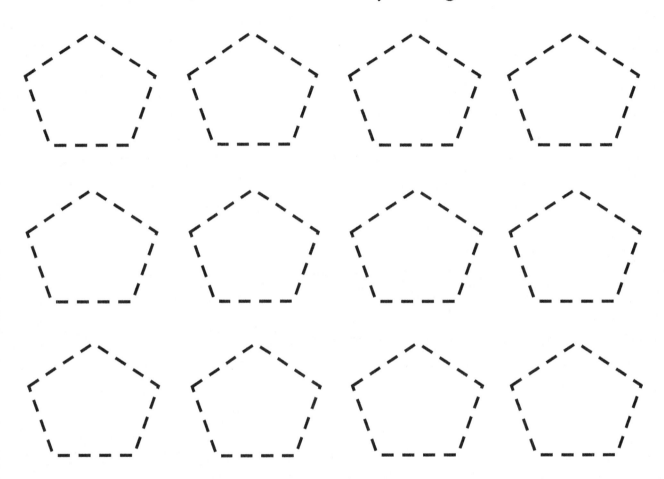

Count how many sides and vertices for pentagon.

| Number of Sides | Number of Vertices |
| --- | --- |
|  |  |

# Geometry

Trace and color cylinders.

Count how many sides and vertices for cylinder.

| Number of Sides | Number of Vertices |
| --- | --- |
|  |  |

Trace and color cubes.

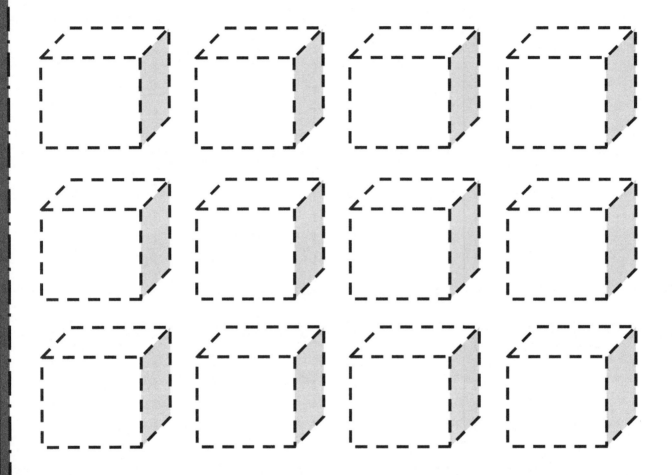

Count how many sides and vertices for cube.

| Number of Sides | Number of Vertices |
|---|---|
|  |  |

## Geometry

Trace and color spheres.

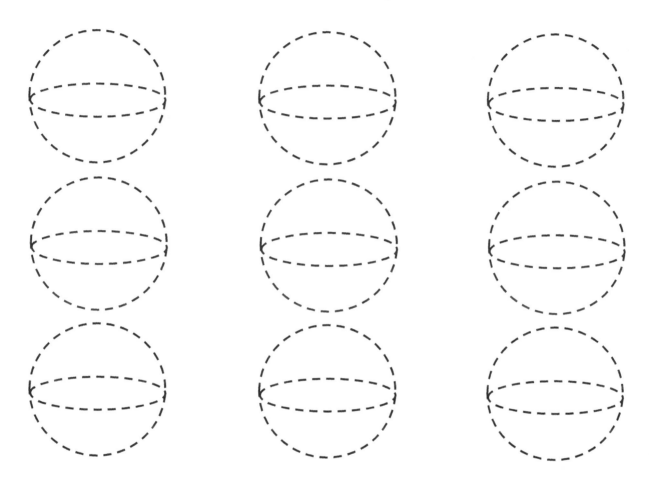

Count how many sides and vertices for sphere.

| Number of Sides | Number of Vertices |
|---|---|
|  |  |

# GAMES AND CHALLENGES FOR SMART KIDS

# Games & Challenges

Use your math skills to find the value of each "?"

TV + TV + radio = 5

radio + radio = 2

radio = [ ? ]     TV = [ ? ]

purse + briefcase = 3

purse + purse + purse = 3

briefcase = [ ? ]     purse = [ ? ]

# Games & Challenges

Use your math skills to find the value of each "?"

🐝 = 7

🦋 = 🐝 + 5

🐝 = 🐜 + 1

🐜 = ⬜ ?

🐝 = ⬜ ?

🦋 = ⬜ ?

# Games & Challenges

Use your math skills to find the value of each "?"

 +  = 6

 +  = 4

 +  = 3

 =  **?**

 = **?**

 = **?**

## Games & Challenges

Use your math skills to find the value of each "?"

🐕 + 🐕 = 20

🦅 + 🐕 = 17

🦅 + 🦅 = 🦊

🐕 = ? 　　🦅 = ?

🦊 = ?

Use your math skills to find the value of each "?"

🚲 + 🚲 + 🚲 = 15

🚲 + ⏰ + ⏰ = 11

⏰ − ☕ = 1

🚲 = [ ? ]

☕ = [ ? ]

⏰ = [ ? ]

Use your math skills to find the value of each "?"

🚢 + 🚢 - ✈️ = 10

✈️ + ✈️ + 2 = 14

✈️ - 🚲 + 🚢 = 11

✈️ = [?]        🚲 = [?]

🚢 = [?]

# Games & Challenges

Use your math skills to find the value of each "?"

🐁 + 🐁 + 🐁 + 🐁 = 16

🐕 + 🐕 + 🐁 = 16

🐕 + 🐰 = 16

🐕 = [ ? ]          🐁 = [ ? ]

🐰 = [ ? ]

Use your math skills to find the value of each "?"

★ + 🌙 = 9

★ - 🌙 = 🌙

🌙 + 🌙 = ★

★ = ☐ ?        🌙 = ☐ ?

You have 3 cows, 2 dogs, and 1 cat.

How many legs do you have?

- - - - - - - - - - - - - - - - - - - - - - - - - - - - - - - - - - - - - - -

How many Triangles are there?

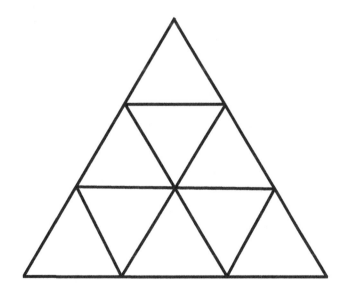

Options:     a) 8  ;  b) 12  ;  c) 10  ;  d) 11  ;  e) 13

# Games & Challenges

Redraw the shape symmetrically.

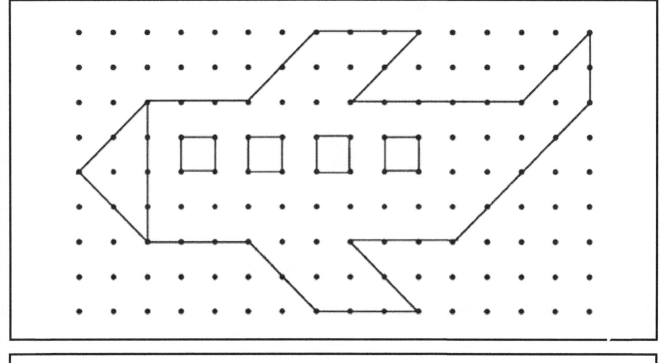

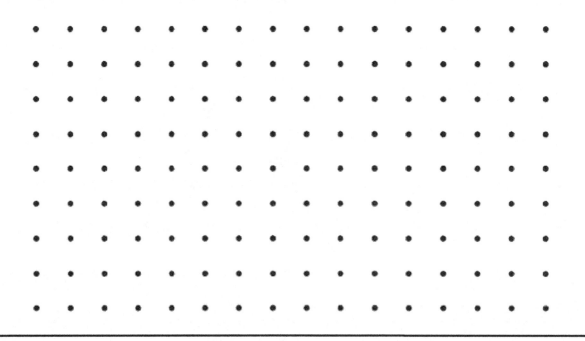

## Games & Challenges

Redraw the shape symmetrically.

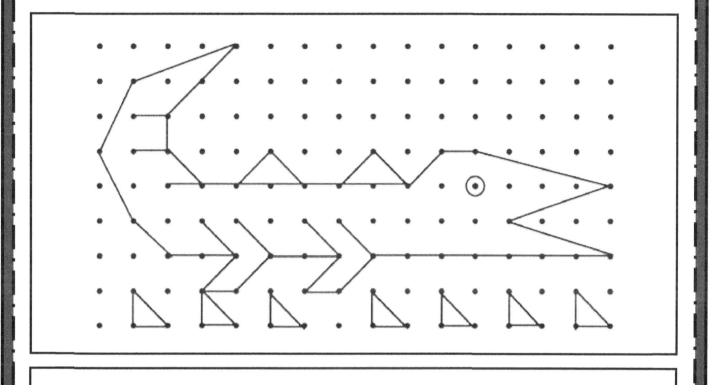

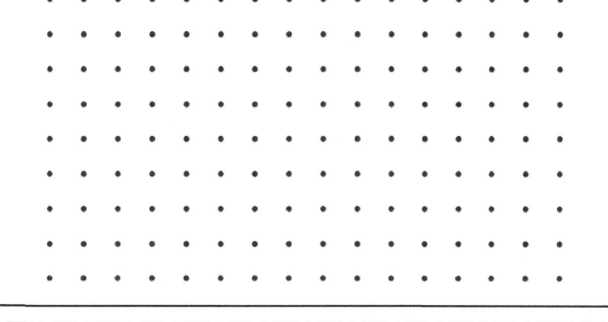

# Games & Challenges

## Redraw the shape symmetrically.

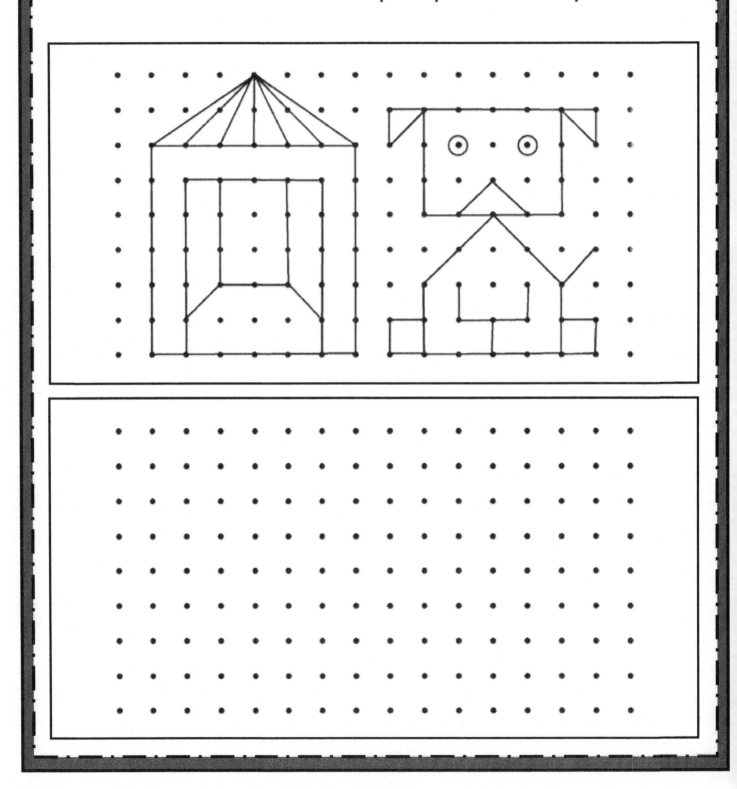

# Games & Challenges

Redraw the shape symmetrically.

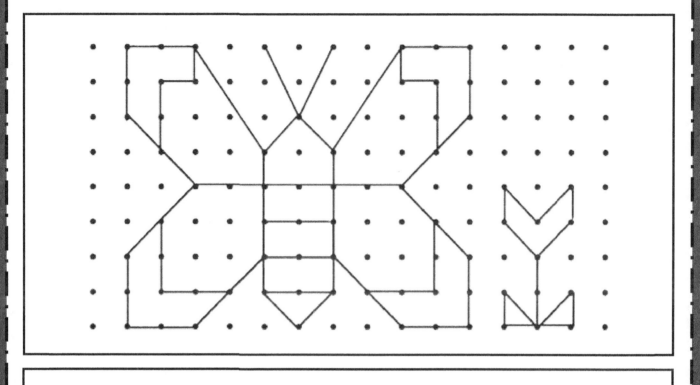

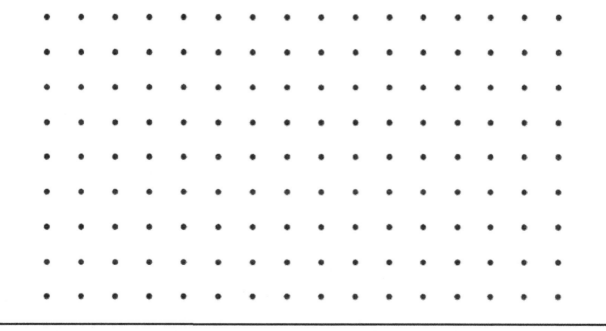

Redraw the shape symmetrically.

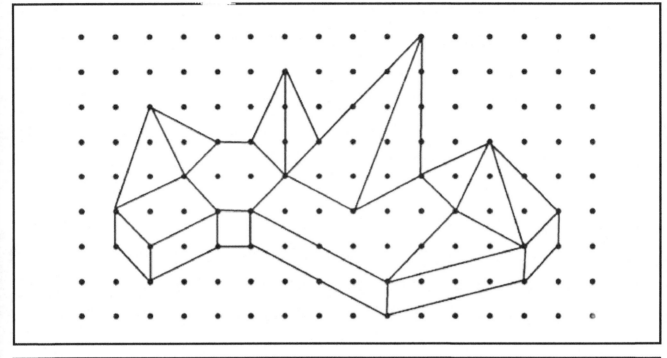

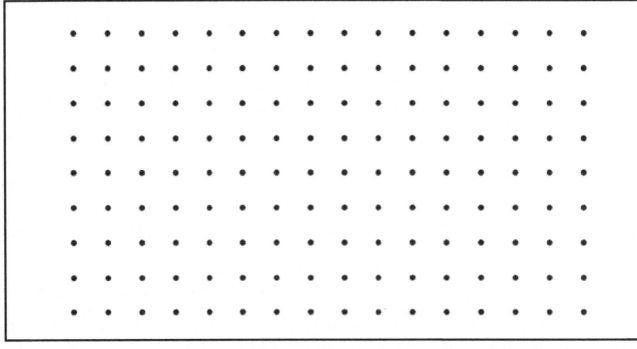

## Games & Challenges

There are two ducks in front of a duck, two ducks behind a duck and a duck in the middle. How many ducks are there?

If five cats can catch five mice in five minutes, how long will it take one cat to catch one mouse?

Susan and Lisa decided to play tennis against each other. They bet $1 on each game they played. Susan won three bets and Lisa won $5. How many games did they play?

You are doing some gardening, and need exactly 4 liters of water to mix up some special formula for your award winning roses.

But you only have a 5-liter and a 3-liter bowl, but do have access to plenty of water.

How would you measure exactly 4 liters?

# Games & Challenges

Each of the integers from 1 to 9 is to be placed in one of the circles in the figure so that the sum of the integers along each side of the figure is 17.

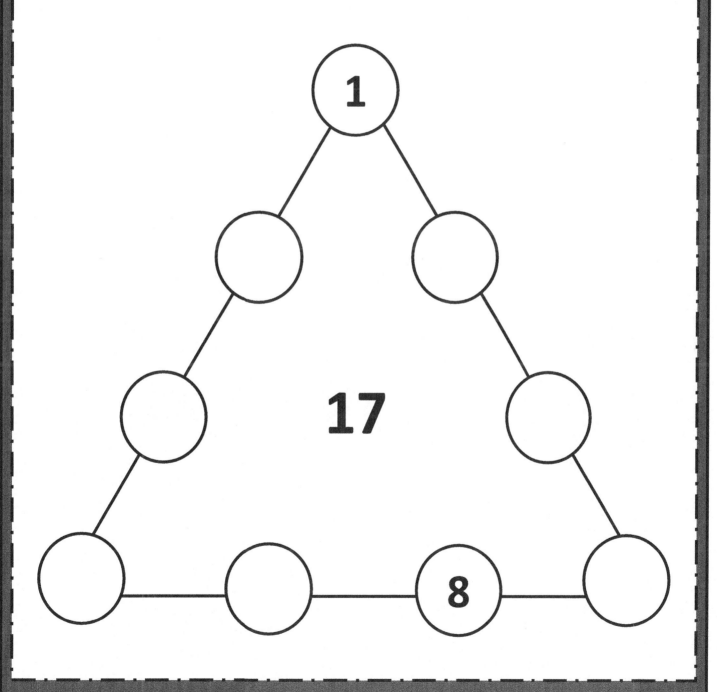

Each of the integers from 1 to 9 is to be placed in one of the circles in the figure so that the sum of the integers along each side of the figure is 19.

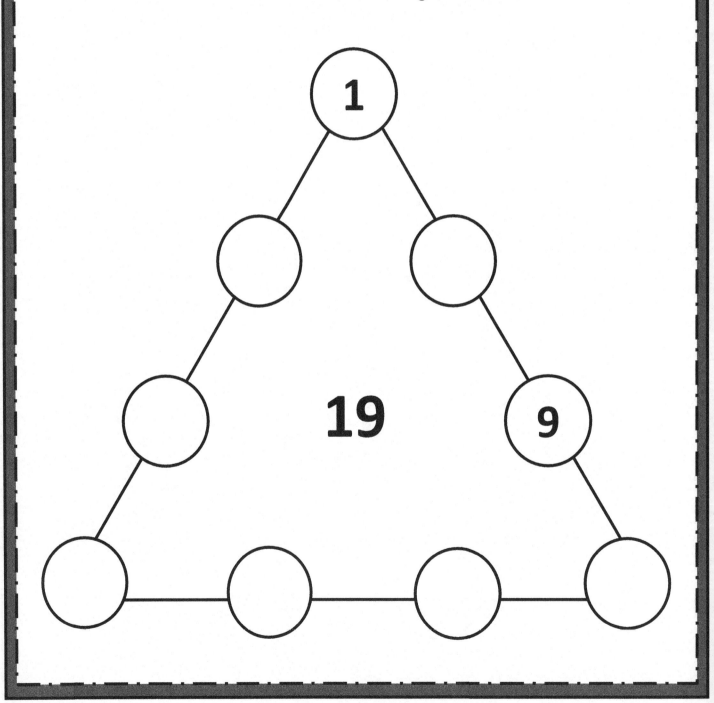

# Games & Challenges

Use your math skills to find the value of each "?"

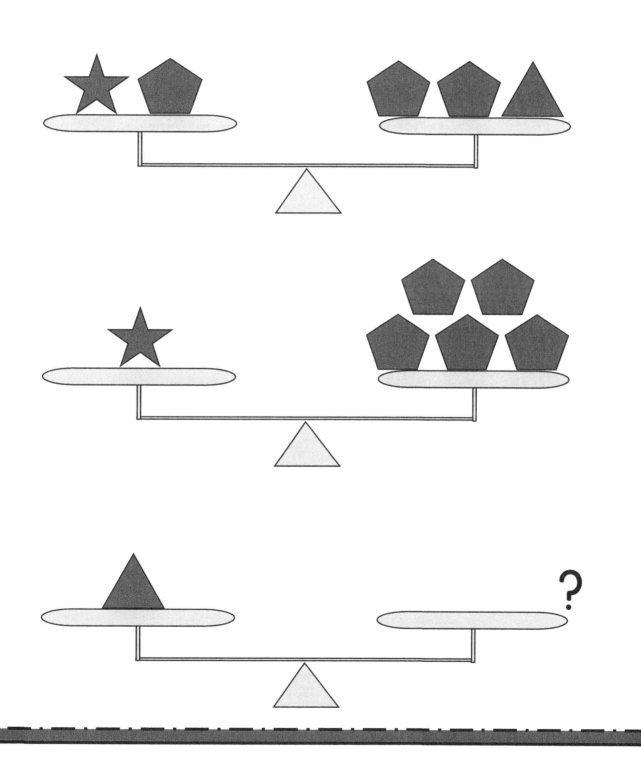

# Games & Challenges

Use your math skills to find the value of each "?"

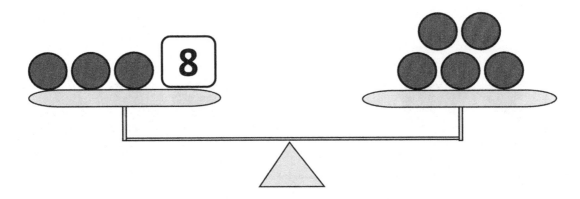

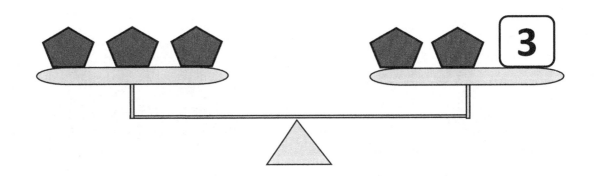

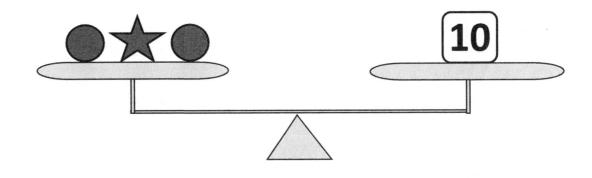

⬤ = [ ] ?    ★ = [ ] ?    ⬠ = [ ] ?

Use your math skills to find the value of each "?"

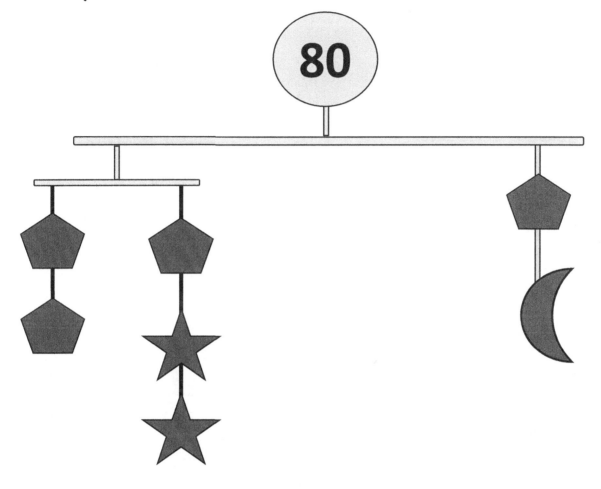

Use your math skills to find the value of each "?"

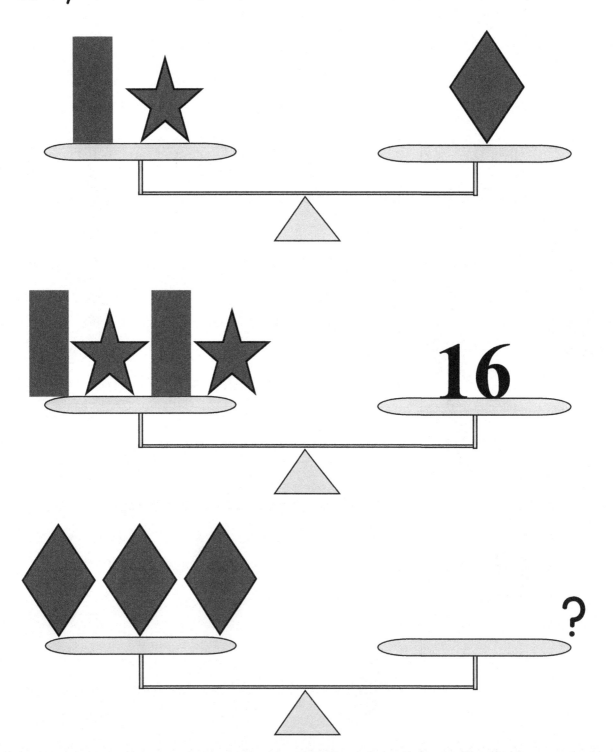

Use your math skills to find the value of each "?"

**24 kg**

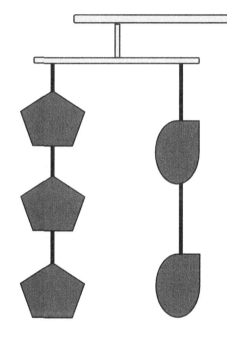
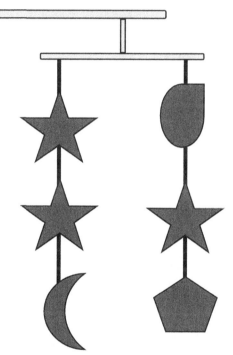

⬟ = [ ? ]

★ = [ ? ]

◗ = [ ? ]

☾ = [ ? ]

Made in the USA
Monee, IL
24 April 2023